AN OCEAN DESERT **THE**

SARGASSO

BY FRANCINE JACOBS **SEA**

ILLUSTRATED BY JEAN ZALLINGER

William Morrow and Company | New York 1975

Printed in the United States of America.
1 2 3 4 5   79 78 77 76 75

Library of Congress Cataloging in Publication Data

Jacobs, Francine.
    The Sargasso Sea.

    SUMMARY: Discusses the location, history, characteristics, marine life, and importance of the unique area of the Atlantic that has spawned numerous unsolved mysteries.
    1. Sargasso Sea—Juvenile literature. [1. Sargasso Sea] I. Zallinger, Jean, illus. II. Title.
GC535.J32   500.9'163'62   74-30376
ISBN 0-688-22029-0
ISBN 0-688-32029-5 lib. bdg.

The author wishes to thank Dr. John Ryther, Senior Scientist, Woods Hole Oceanographic Institution, for reading and checking the manuscript of this book.

CONTENTS

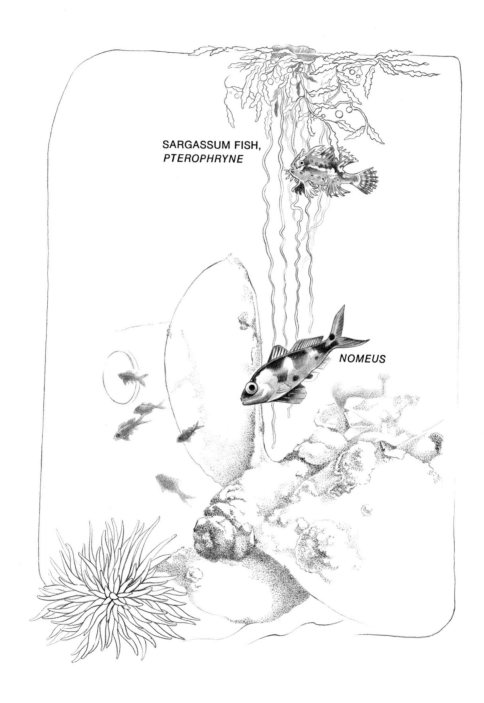

SARGASSUM FISH,
PTEROPHRYNE

NOMEUS

I   THE GRAVEYARD
OF THE SEA

Ever since ancient seafarers dared to enter its
weedy waters, the Sargasso Sea in the North
Atlantic Ocean has been shrouded in myth and
mystery. Even today there is an air of
strangeness and wonder about it. Ships and
planes bearing altogether some three hundred
persons in this century alone have mysteriously disappeared in the western Sargasso Sea,

in an area that has become known as the Devil's Triangle.

Sailors have told such terrifying and incredible tales about the Sargasso Sea that many people doubt its existence, except in legend. There are stories about vast, sweeping masses of weeds, which float over hundreds of miles of ocean. Seaweed was supposed to be so thick and dense in places that it trapped sailing ships, even large steamers. Unending heaps of it have been described growing over bows, climbing masts, crushing hulls, and choking propellers. The seaweed is said to be so strong that its long, entangling tendrils have held vessels even in the fiercest wind.

There are legends too of monsters. Terrible sea serpents of gigantic size were thought to live in the Sargasso. They were believed to lurk beneath the weed, waiting to attack any unsuspecting vessel that came near. Stories had them smashing hulls and wrapping them-

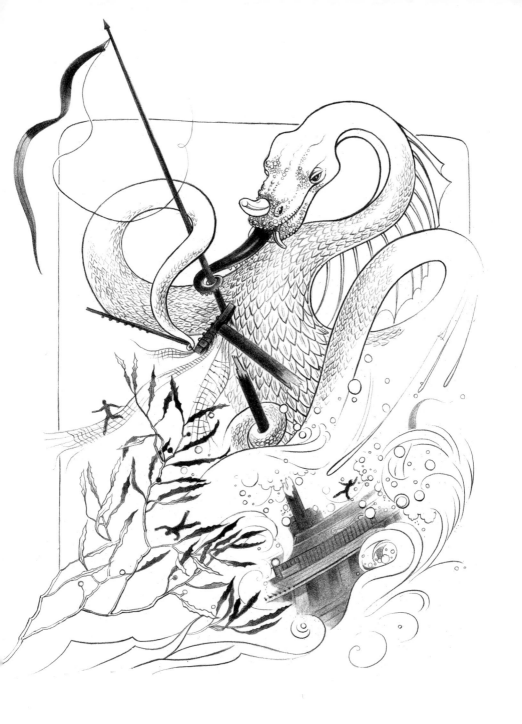

selves about helpless ships, dragging them with their sailors down into the sea.

There are tales also of great circling currents called "eddies," so powerful that they pulled ships vast distances within the Sargasso. These currents were said to draw vessels, wrecks, and debris into the Sargasso's calm and windless center. Here, according to fable, they remain exhibits in a strange museum of captive ships circling endlessly about. Withered hulks of an-

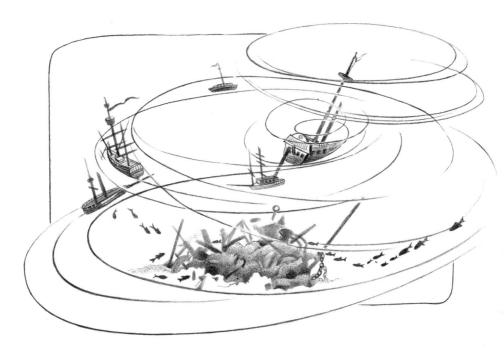

cient Phoenician galleys float at the center, surrounded by the rotted shells of galleons, brigantines, and caravels, flying the tattered pennants of old Spain, France, and England. Pirate ships, their crow's nests empty, their decks deserted, their plunder untouched, drift about. Beyond them circle clipper ships, cruisers, and rusting steamers in eerie silence. With such legends, is it surprising that the Sargasso Sea is known as the Port of Missing Ships and the Graveyard of the Sea?

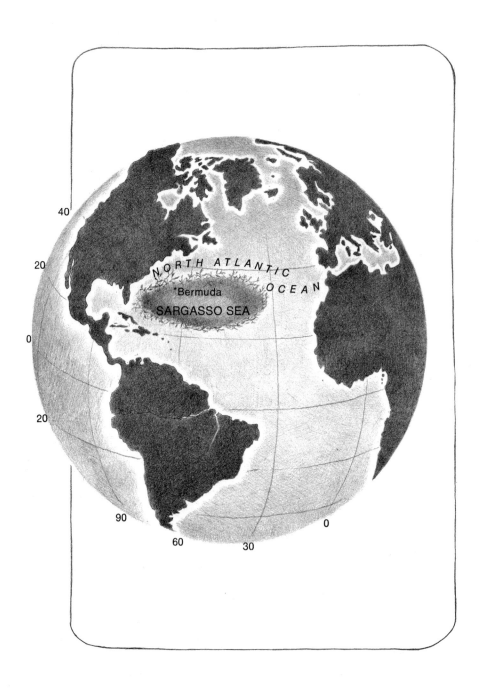

II DISCOVERY

Despite its legendary character and awesome reputation, the Sargasso Sea is a reality. It exists. It is a weed-strewn, oval-shaped body of water lying in the ocean a thousand miles from any mainland. On a map or globe, it is shown stretching from the middle North Atlantic toward the southeastern coast of the United States, between longitudes of thirty

PHOENICIAN COINS AND TRADE ITEMS

degrees to seventy degrees west and latitudes of forty degrees to twenty degrees north. It surrounds the Bermuda Islands, which lie northwest of its center.

Who were the first people to venture into this mysterious part of the Atlantic? Were they responsible for the Sargasso's legends? Many scholars believe the ancient Phoenicians discovered the Sargasso Sea and deliberately invented frightening tales about it. The Phoenicians were a hardy, seafaring people, who lived some three thousand years ago on the shores of the Mediterranean Sea. They were traders who built sturdy sailing vessels called galleys. With courage, a good knowledge of astronomy, and the stars to guide them, they are said to have traveled long distances in order to trade. It is believed that they sailed all around Africa and as far north as the British Isles. Phoenician coins in a jug that washed up onto an island in the Azores near the Sargasso Sea suggest that the Phoenicians had been

there as far back as the fifth century B.C.
Perhaps they had been trading with people living in the Azores then, or perhaps they were
merely passing through the Sargasso Sea on
their way to the Americas and were
shipwrecked.

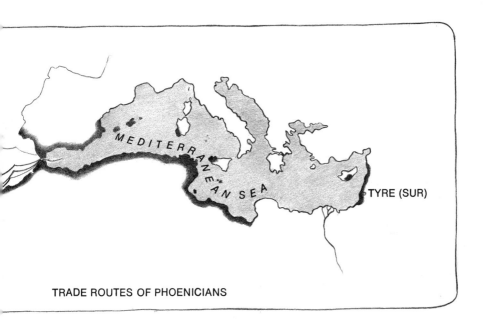

MEDITERRANEAN SEA

TYRE (SUR)

TRADE ROUTES OF PHOENICIANS

The Phoenicians were cunning, and they kept their trade voyages and routes secret. They were even known to mislead their trading rivals, the ancient Greeks and Romans, purposely. To discourage their competitors then, the Phoenicians may well have originated frightening tales of the Sargasso. If so, their strategy probably succeeded. Few areas on earth have so dreadful a reputation.

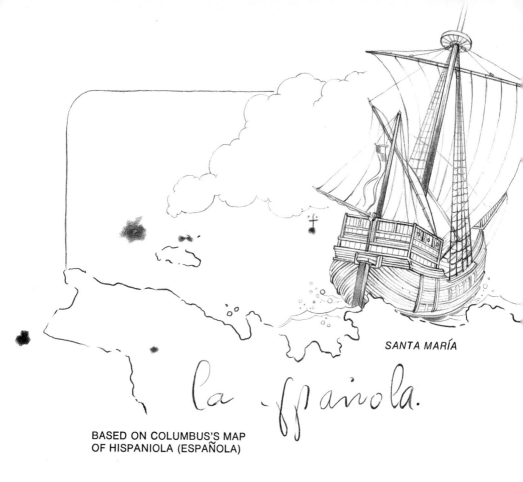

SANTA MARÍA

La ḟpañola.

BASED ON COLUMBUS'S MAP
OF HISPANIOLA (ESPAÑOLA)

For the first recorded description of this sea, we credit Christopher Columbus. In mid-September in 1492, during his first voyage across the Atlantic, Columbus reached a point some eight hundred miles west of the Canary Islands. He was on the edge of the Sargasso Sea.

18

Entries in his ship's log record the sea covered with green and yellow weeds and note the presence of a crab among the weeds. Believing that seaweed and crabs existed only along shorelines, the sea-weary sailors pressed onward searching for land. As days went by and the weeds increased, the crew became alarmed that their course would be blocked. At first Columbus was not unduly concerned. He continued to search for islands but without success.

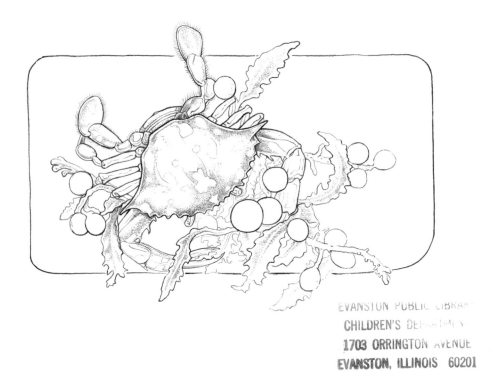

Still in the Sargasso Sea, the crew became increasingly frightened and restless, and Columbus tried to calm them. He even kept a separate log in which he made false entries to suggest to them that they had not really gone so far from home. Columbus ordered a sounding, the lowering of a weighted line to measure the sea depth. He was convinced, in view of all the weeds, that the sea was shallow and that land was close by. The sounding line was one hundred fathoms, or six hundred feet, long. This length was generally more than adequate to reach the bottom of the coastal waters of Spain, Portugal, and Africa, where these seamen usually sailed. But the line didn't reach the bottom of the Sargasso Sea. They joined two lines together. At twelve hundred feet the bottom was still beyond reach. The great depth worried the crew but not nearly as much as what happened next.

The expedition was becalmed. The winds,

upon which their sails depended, all but disappeared. Without knowing it, the voyagers had reached the center of the Sargasso Sea, an area of long calms. Here upon the almost motionless sea, the caravels sat for some days, making very little progress. "The sea was smooth as a River," the log reveals. Fear was widespread. If the calms continued, their provisions of food and water would run out. The crew was on the verge of mutiny and panic. They worried that there was not even enough wind to take them homeward to Spain. At last Columbus's courage faltered.

After drifting very slowly for some days, the voyagers wandered once more into the trade winds and their journey westward resumed. The seaweed thinned and land was soon sighted. Columbus had not found the water route to Asia that he sought. Instead, he discovered the New World. On his difficult journey, he had, also without intending it, provided the first authentic description of the Sargasso Sea. He recorded observations of the weed with small, scattered air-filled pods and brightish green tips projecting upward from deeper brownish masses.

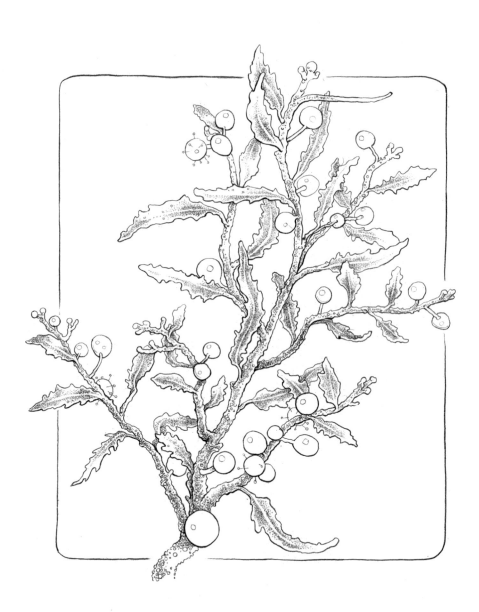

This weed is called "sargassum." It has jagged edges and resembles holly. Unlike other closely related forms of sargassum, which are rooted and grow in coastal areas, sargassum that lives in the Sargasso Sea is not attached to the ocean floor. It floats. Small air-filled bladders, which act like tiny buoys, keep it on the surface. Columbus termed these pods *"coma fruta,"* or "sort of fruit." Portuguese seamen who followed in Columbus's path called them *"salgazo"* after grapes. These weedy waters, or sea of grapes, thus became the Sargasso Sea.

Though almost five hundred years have passed since Columbus safely journeyed through the Sargasso, tall tales and fantastic stories about the weeds persist. Explorers, conquistadors in search of gold, colonists flee-ing persecution, and pirates may all have had reasons, like the old Phoenicians, to exagger-ate and invent dangers in order to discourage pursuit. These tales were so accepted that even as recently as 1952 a French physician crossing the Atlantic in a rubber boat purposely avoided the Sargasso Sea. "The whole area has always been a major navigational hazard," he wrote, "a terrible trap where plant filaments and seaweed grip vessels in an unbreakable net."

Old sailors' stories, paintings, novels, and motion-picture dramas continue the fabled reputation of the Sargasso. The truly remarkable and unusual nature of this celebrated area, however, is less well known than its legends. Its unique character and its strange, unsolved mystery are not myths.

III   LOCATION

You might think that defining the location of a sea would be easy. Usually one would rely upon its land borders. Thus, the Mediterranean Sea could be described as lying between Europe and North Africa. But the Sargasso Sea has no land boundaries.

In the nineteenth century, two German scientists took up the puzzling question of the

Sargasso's location. It was not an easy task. Alexander von Humboldt relied on what was known about Atlantic currents in the early 1800's to define the Sargasso's borders. It is a "caged" sea, he said, a prisoner of the surface currents that swirl around it in the North Atlantic. He believed the course of these currents to be constant and that the Sargasso was always in the same place. This assumption proved to be untrue.

Otto Krümmel, later in the 1800's, tried to locate the Sargasso by noting the limits of the seaweed. He methodically recorded sightings of the weeds from the logs of German merchant ships crossing the Atlantic. Placing this information on maps, he was able to outline the Sargasso Sea, much as you would join dots with lines to form a picture. The picture Krümmel made showed the Sargasso to be an oval-shaped area stretching from the middle Atlantic toward North America.

Similar efforts in the 1920's, by the Danish botanist, Øjvind Winge, enlarged the Sargasso's borders. They were expanded eastward into the central Atlantic and southward to the islands of the West Indies, south and east of Florida.

In the 1930's and 1940's, oceanography, the science of the sea, advanced rapidly. The world-famous oceanographic center at Woods Hole, Massachusetts, came into existence. Marine scientists at Woods Hole began intensive studies of the Sargasso Sea, studies that are still in progress today. Much of their early research was done aboard the *Atlantis,* a ship that was virtually a floating laboratory. Supplied with modern equipment, scientists began an analysis of the Sargasso Sea that would help solve the problem of its location and lead to an astounding discovery.

ATLANTIS

Hundreds of Nansen bottles, designed to gather water samples at specific depths and to record temperatures at those levels, were lowered into the sea. These samples were brought aboard ship for study. The Sargasso was found to be warmer and saltier than the ocean around it. The Atlantic is the saltiest ocean on earth, and the Sargasso Sea proved to be the saltiest part of the Atlantic. It is so salty that by taste you can easily tell the difference between it and seawater from Cape Cod, Massachusetts, for example. Its unusual saltiness, or salinity, and its warmth gave scientists a reliable way to determine the borders of the Sargasso Sea.

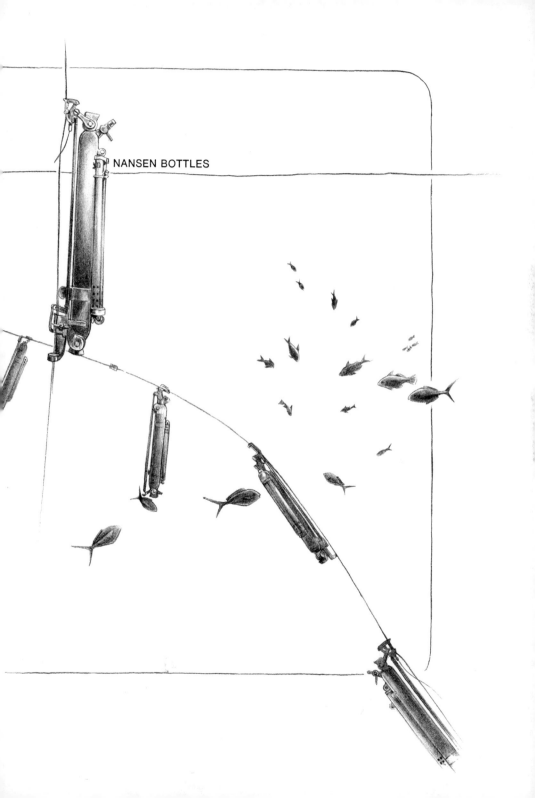

NANSEN BOTTLES

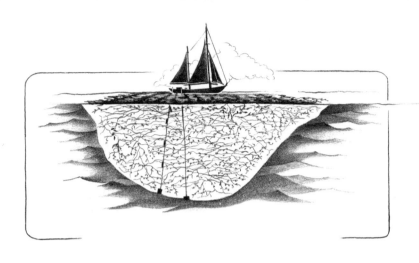

Oceanographers also learned that this warm, salty sea is relatively shallow. Even though the ocean below it is very cold and very deep, reaching down about four miles, the Sargasso Sea descends little more than half a mile at its center, where it is deepest. The Sargasso is shaped like a peach pit, thicker in the middle and tapered toward the edges. It is an unusual sea that floats upon the ocean much as oil floats on water. Though the floating weed may overflow its shoreless borders, measurements of temperature and salinity will always mark the Sargasso's actual boundaries.

Of the Sargasso Sea, Rachel Carson, the world-famous biologist, wrote: "It is so different from any other place on earth that it may well be considered a definite geographic region." What accounts for its uniqueness? Scientists believe that because it is so far from land, the Sargasso fails to receive fresh water from continental rivers. Without these inflows it remains relatively undiluted and highly salty. In addition, much water evaporates at the surface leaving salt behind. There is little rain here to replace the lost water, which also contributes to the Sargasso's saltiness.

The temperature of the Sargasso Sea is determined by the warm surface currents that surround and contribute to it. The sea resembles the hub of a slowly turning wheel. The southern rim of this great wheel is the North Equatorial Drift, which parallels the equator. To the west, the rim is formed by the powerful Gulf Stream, and to the north, by the North Atlantic

Current. The Canaries Current on the east completes the wheel, though actually the Sargasso's eastern boundary conforms more precisely to the Mid-Atlantic Ridge. This long underwater mountain range running north and south splits the Atlantic bottom into two huge basins.

The Sargasso Sea's position shifts, because these prevailing surface currents wander and at times even peter out. The Canaries Current may weaken to the point where it is hardly a barrier at all. The North Equatorial Drift shifts with the seasons. The Gulf Stream's path continually changes. In a month's time, it can

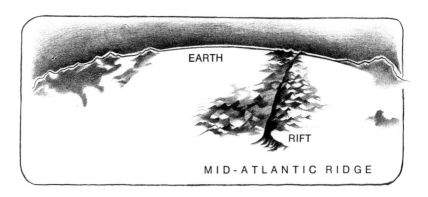

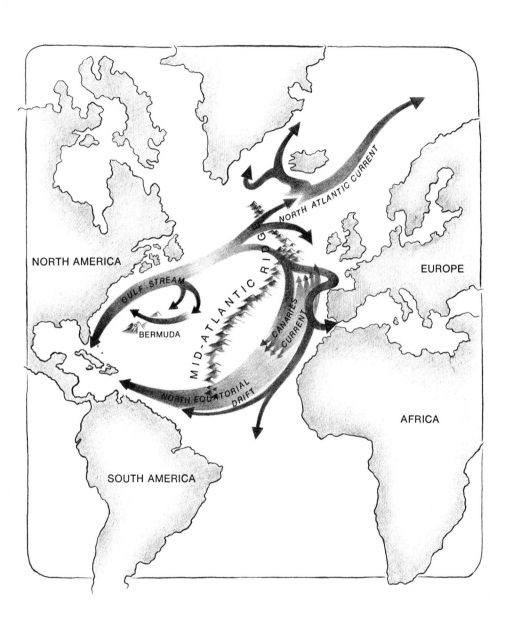

move as much as one hundred miles from its previous course. The North Atlantic Current splits apart forming vast circling eddies. These great whirlpools may be an astonishing two hundred fifty miles long. They siphon millions of tons of water into the Sargasso Sea.

The vast ring of currents that surrounds the Sargasso Sea is called a "gyre." This gyre moves in a clockwise direction. Though the Sargasso Sea is some two million square miles in area, greater than half the size of the United States, its volume depends on these constantly varying currents. Difficult as it is to imagine, these currents may move as much as seventy-five million tons of water each second. Their circular motion and force, aided by the earth's rotation, cause water to heap up toward the Sargasso's center. For this reason the sea is thicker at the middle and tapered toward the ends. It is as much as two feet higher in the center than at the edges.

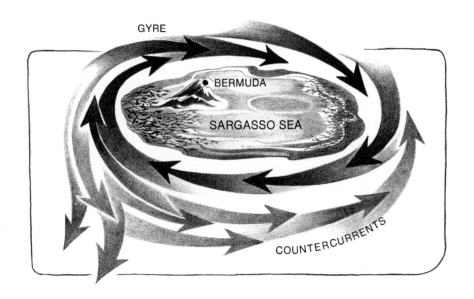

GYRE

BERMUDA

SARGASSO SEA

COUNTERCURRENTS

Once it was believed that the ocean beneath the surface was a motionless place. This supposition is not so. Currents travel not only on the surface but throughout the depths as well. Sometimes one current may even travel below another, moving in the opposite direction. It is called a "countercurrent." A deep, cold, slow countercurrent moves southward beneath the warm Gulf Stream on the western edge of the Sargasso Sea. The depths are anything but calm.

Great deep undersea currents are constantly creeping along throughout the world's oceans. The depths beneath the Sargasso Sea contain, among others, cold currents from the distant Antarctic and Arctic Oceans and warmer, salty ones from the Mediterranean Sea. Each current takes its place, one above the other, like the layers in a cake. You might think of the Sargasso Sea as the icing on this cake, for it differs greatly from the layers beneath it.

IV   AN OCEAN DESERT

The Sargasso's most striking difference is its
unique transparency. Its waters are almost
crystal clear. They are deep blue and so per-
fectly transparent that a large yellow disk
lowered on a rope from a research ship was
visible to the crew on deck to a depth of 217
feet. This distance is as much as the height of a
twenty-story apartment building. On sunny

41

days, descending light rays have been recorded on photographic plates at 3250 feet. This is an extraordinary depth for light rays to penetrate and shows how remarkably transparent these waters are. Why is the Sargasso so clear? Plankton, microscopic plants and the tiny animals that feed upon them, which are numerous elsewhere and give other seas their murky green color, are sparse here.

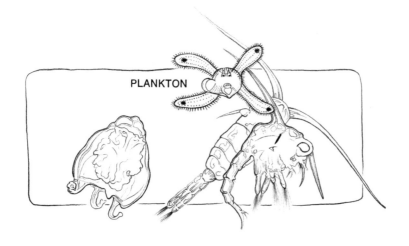

PLANKTON

The Sargasso is so remote that it does not receive the fertilizing nutrients normally washed into coastal waters by continental rivers. Moreover, few upwellings occur here. These risings of deeper waters, bearing nitrogen, phosphorous, and other minerals needed by plants, are produced elsewhere in the ocean by winds and opposing currents. In the Sargasso Sea, the stable body of warm surface water, which sits on the colder, deeper waters below, resists such vertical mixing. As a result, the Sargasso Sea is not as fertile as other ocean areas. There is little to support the growth of plankton.

GALATHEA

In the early 1950's, researchers aboard the Danish vessel, the *Galathea,* sampled sea-water in all the great oceans of the world. The Sargasso Sea was discovered to have the least amount of plankton. It was found to produce only about one third as much as the average found elsewhere. Although it has some marine life, when compared to other ocean areas, the Sargasso Sea is relatively unfertile. It is a desert in the sea, a great ocean desert.

It is strange at first to think of a desert in the sea. Yet the Sargasso has a definite relationship to other great desert areas in the world. You can discover this connection yourself with a map or a globe. Place your finger on the Sargasso Sea. Using your finger as a guide, proceed westward, roughly between twenty degrees and forty degrees north latitude. Crossing the United States, you come to the great deserts of the Southwest and of northern

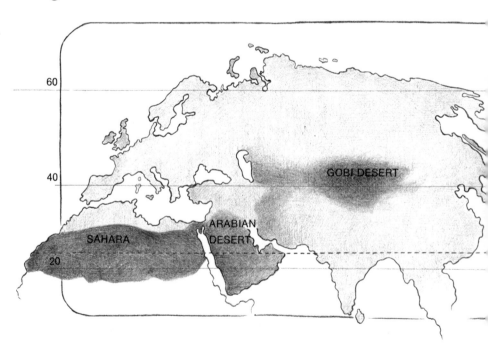

Mexico. Across the Pacific and onto the land mass of Asia, you pass through the great Gobi Desert of China, the deserts of India, Pakistan, and Arabia. On into Africa, you go through the desert areas of Egypt, Libya, Algeria, and Morocco — the vast, arid Sahara. Moving westward again, crossing the Atlantic you arrive once more back where you began in the Sargasso Sea. The Sargasso and other great deserts lie in a dry belt that circles the earth.

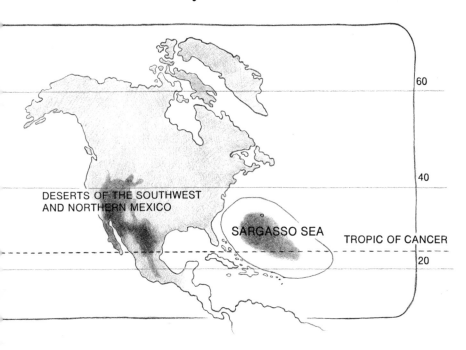

Your map or globe may also indicate that the Sargasso Sea is found in the horse latitudes at thirty degrees north, the widest part of the North Atlantic. This area is a zone untouched by the northeast trade winds that steadily blow westward from Europe toward the equator and across the Atlantic. Columbus was becalmed here. So were many of the ships that later followed in his path to the New World. It is said that horses aboard Spanish vessels died of thirst here and had to be dumped overboard. Thus, so the story goes, the name horse latitudes came to describe this windless zone, which passes through the ocean desert.

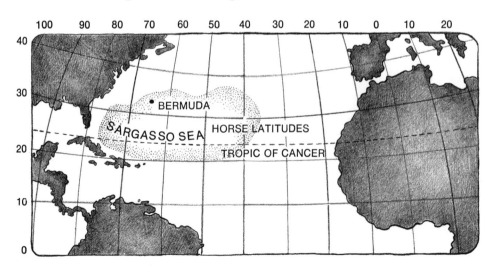

The weed that floats upon the Sargasso is an
oasis in the desert sea. It is found in bunches
like tiny islands and in masses as large as a
football field. But the legendary weedy jungles
of myth do not exist. While lumps of sargassum may be several feet thick, most of the
weed, perhaps as much as seven million tons,
is thinly spread, like scattered hay, throughout
this vast sea.

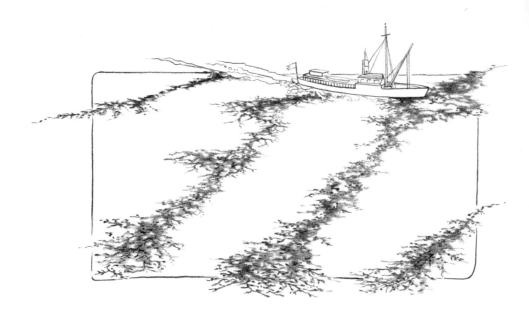

Prevailing winds and currents often push the floating weeds into long lines. These lines are called "windrows." Sometimes the windrows are twenty feet wide and extend many miles, even from horizon to horizon. Between these weed lines are lanes of clear, blue water, and the Sargasso Sea at such times looks like an olive-and-blue-striped carpet. At other times, very infrequently, the weed mysteriously is almost absent.

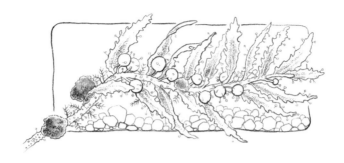

Tiny mosslike creatures can attach themselves to the weed, and sometimes they so heavily infest sargassum that it sinks. If a plant escapes them and is not broken by a storm or blown away into colder waters, it is capable of an extremely long life. Doctor Albert Parr, who spent years studying this weed, believes that in the calm, still waters at the center of the Sargasso Sea, plants may live for many decades. In fact, certain individual plants are virtually immortal; they may live for centuries. "It may well be that some of the very weeds you would see if you visited the place today were seen by Columbus and his men," wrote Rachel Carson.

Sargassum is the only vegetation not micro-scopic in size to live its entire life as an in-dependent, free-floating plant in the open sea. It is so adapted to a floating life that it lacks even the rootlike "holdfasts" other seaweeds use to anchor themselves. It is a simple plant without true leaves, stems, and roots. Its leaf-like blades called "fronds" are frail and thin. Although it is yellow-brown in color, sargas-sum does contain chlorophyll, the green chem-ical that enables plants to make their own food. There are some eight similar species of the weed found in the Sargasso Sea, though only two kinds predominate.

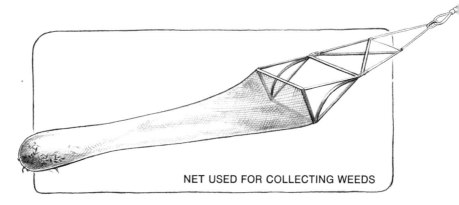

NET USED FOR COLLECTING WEEDS

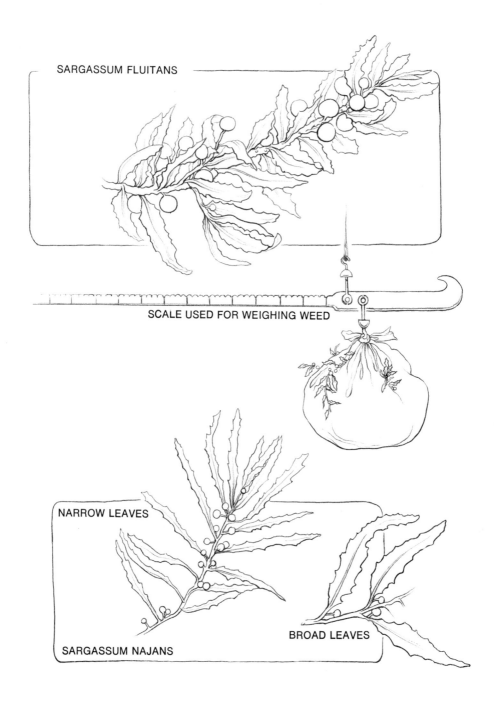

SARGASSUM FLUITANS

SCALE USED FOR WEIGHING WEED

NARROW LEAVES

SARGASSUM NAJANS

BROAD LEAVES

Sargassum belongs to one of the oldest families of plants on earth, the algae. Its ancestors were the tiny primitive plants that once covered ancient seas two or more billion years ago. Sargassum found in the Sargasso Sea has neither special reproductive cells nor spores common to other seaweeds. Its means of reproduction long baffled scientists. Many believed that it really did not reproduce at all but merely gathered here, having drifted from Floridian and West Indian shores. Now it is known, however, that the sargassum found here is a unique, self-perpetuating resident of the Sargasso Sea. It reproduces vegetatively by very simple means. As fragments of the weed break off they grow, enabling this form of life to continue here in the ocean desert.

V   THE WEED COMMUNITY

Just as an oasis supports life in a land desert, so the sargassum weed serves as a special outpost of life in this relatively barren sea. The weed is host to a diverse community of permanent, part-time, and visiting creatures. It offers its inhabitants food, shelter, and a protected environment for their young, rare advantages out on the surface of the open sea. Living here

is an array of animals in many ways more strange and fascinating than the mythical monsters once imagined.

Most of these unusual creatures are small. Some are no larger than the head of a pin. To explore a clump of sargassum with a magnifying glass can be an exciting experience. If you should ever visit Bermuda in the Sargasso Sea, try to collect the weed that drifts offshore. Examine it closely, for the creatures that live here are unusually well hidden and disguised. Shake it over a pan of water, and you may dislodge some amazing hitchhikers.

56

There are animals without backbones, animals with backbones, cannibals eating their own kind, and predators that prey upon others. The wingless water strider, *Halobates,* the only insect found in the ocean, uses the weed as a raft on which it strides about with its spindly legs searching for food. Some fifty-

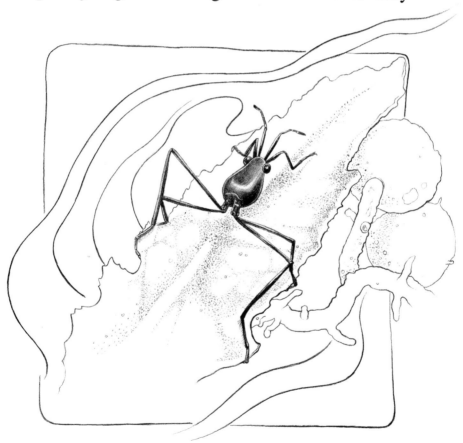

FLYING FISH

PUFFER

four species of fish are known to associate with
the weed in some way. The strange, spiny,
inflatable puffer is common. Jacks, trigger-
fishes, and various filefishes, whose rough
body feels like sandpaper, abound. Marlin,
swordfish, and sailfish, the great Atlantic bill-
fishes, spawn here. The graceful flying fish
glides round and round, binding sargassum
together, building nests into which strings of
its eggs are deposited. Hatchling flying fish are
born brownish yellow, the color of the protect-
ing weed. When they are older and ready to
leave this nursery in the ocean, they turn silvery blue like the shimmering sea.

MARLIN

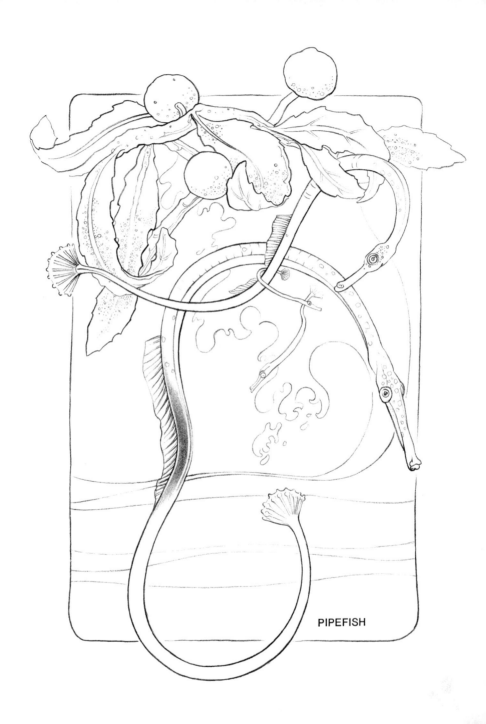

PIPEFISH

Also taking shelter in the sargassum are slender finger-length fishes that resemble needles. They are the pipefishes, and they have long, toothless, tube-shaped snouts for sucking in food. They not only look like branches of the weed, they swim gracefully in rhythm with the waving fronds and are easily mistaken for them. Closely related to the pipefish is the sea horse. Covered with bony plates of armor, it seems more like a reptile than the fish that it truly is. The sea horse hides here holding onto the weed by curling its tail around it. Like its cousin, the male pipefish, the male sea horses, strangely, also bear the young. That is, they carry the eggs deposited by the female in a pouch below their stomach for some six weeks until the babies are hatched out. Sea horses in the Sargasso are experts in disguise, matching themselves to the yellow weed in daytime and turning black at night.

The most unusual creature in the Sargasso

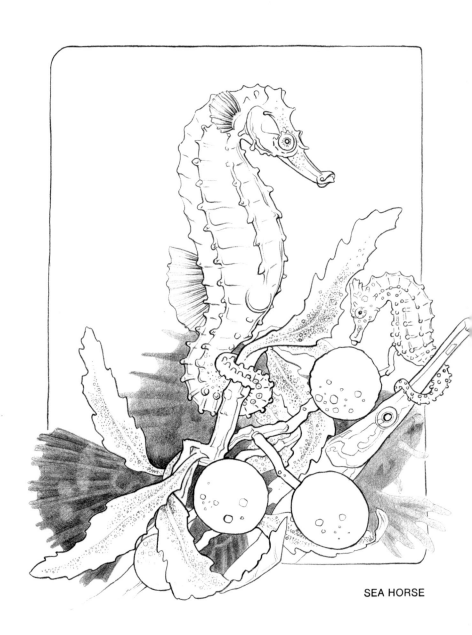

SEA HORSE

Sea is perhaps one of the strangest backboned animals on earth. Its scientific name is *Pterophryne* (pronounced Ter'-o-fry'-nee), but more commonly it is known as the sargassum fish, for it spends its entire life in sargassum. A tiny but ferocious predator, it grows to about six inches and is magnificently camouflaged. Not only does it have the weed's color, it has body markings that imitate the weed's round air bladders and branching fronds. These extraordinary markings continue on into its eyes. All over the skin there are tiny white dots matching the homes of tube worms that live upon the weed. Leafy appendages grow from its tiny body, giving *Pterophryne* an irregular outline, like a sprig of weed. As further camouflage, the sargassum fish also alters its color, like a chameleon, when its background changes.

What makes *Pterophryne* so queer a fish is its odd front fins. These pectoral fins found

behind the head are stiff and rigid in most
fishes, but those of the sargassum fish are
jointed like hands. With these peculiar
"hands," *Pterophryne* grips the weed. In this
miniature jungle, where wars for survival are

64

UNDERSEA CAMERA FOCUSSED
ON THE SARGASSUM FISH, *PTEROPHRYNE*

constant, this monstrous little cannibal swings
from frond to frond like a monkey. For its
small size, it has a huge mouth with many

teeth. A fleshy lure suspended from its head dangles before this dangerous trap. Other fish, shrimp, worms, and even other sargassum fish of almost equal size are caught. When attacked by larger predators, *Pterophryne* snaps open its enormous mouth, gulps down quantities of water, and swells to such a size that the attacker must either cough it out or choke.

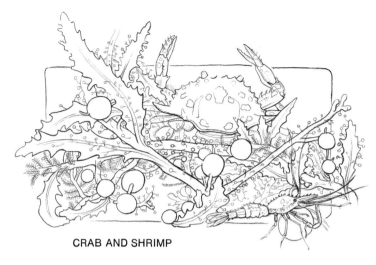

CRAB AND SHRIMP

Food for the fishes here is provided by many creatures, especially by the reddish and yellow-brown shrimp and crabs. These crabs have the remarkable ability to camouflage themselves in the weed. Snails live here, too, drilling tubular holes in the weed, while certain worms crawl about hollowing sargassum stems and fronds. Other worms dwell in tiny, white, spiral-shaped tubes and poke their head out to gather food. Tiny flowerlike hydroids decorate the weed. These little animals, belonging to the jellyfish group, live together in colonies attached to the weed, making the sargassum appear fringed.

Up and down the weed crawls one of the sargassum's most unlikely monsters. It is the sea slug, a soft, brown-bodied, shell-less member of the snail family. Another master of disguise, this tiny creature is hard to find. Its weed-colored body has folds and flaps that change and conceal its outline as well as

SEA SLUG, *SCYLLAEA PELAGICA*

circular markings that copy the pods. This squishy, harmless-looking little fiend bears the well-earned scientific name, *Scyllaea pelagica.* Scylla was a mythical sea monster, a sort of many-headed octopus that gobbled down passing sailors. The sea slug devilishly engulfs innocent passersby too, forcing them down its mouth and along its gizzard with spearlike spines. When it wishes, *Scyllaea* can also swim from one clump of weed to another.

Among the weeds and in the clear waters of the Sargasso there lives an odd couple, *Physalia* and *Nomeus. Physalia* is a soft-bodied relative of the jellyfish and is better known as the Portuguese man-of-war. A drifter, whose gas-filled float may be pink, blue, or purple, it is about a foot long and relies on its delicate sail and the currents to get about. *Nomeus,* its partner, is a finger-long, silvery and blue-black, striped fish and is such a close companion to *Physalia* that it is also com-

PHYSALIA

NOMEUS

monly known as the man-of-war fish. This relationship, however, is more than mere friendship. Each creature helps and receives benefit from the other.

Nomeus makes its home among the man-of-war's dangling tentacles, an unlikely refuge since these stinging streamers can inject poison nearly as deadly as a cobra's. Even human beings unfortunate enough to swim into or to touch them have suffered intense pain, illness, or death. *Nomeus* gracefully avoids contact and may even have a degree of immunity to this poison. Its vertical stripes and coloring conceal *Nomeus* among the tentacles, but should predators such as other fish attack, *Nomeus* leads them into a deadly trap. *Physalia's* stinging, stretchable tentacles, which can be more than fifty feet long, paralyze the intruder and in seconds hoist it up to the float. Here it is held and digested while *Nomeus* enjoys the leftovers.

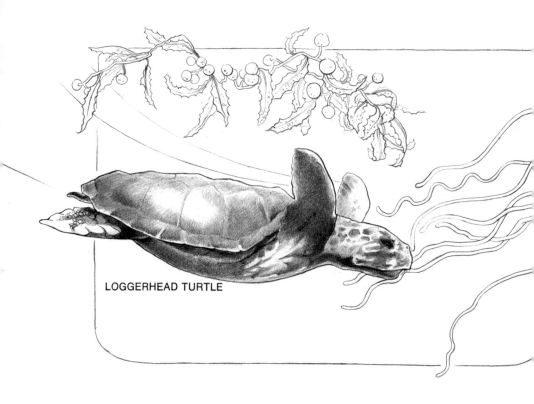

LOGGERHEAD TURTLE

The giant loggerhead sea turtle, however, relishes the men-of-war, gobbling them up tentacles and all with no apparent ill effects. Young sea turtles, whose whereabouts in the sea is still one of nature's mysteries, may be hiding out in the Sargasso. That is the theory of Professor Archie Carr, a renowned zoologist and expert on sea turtles. Carr thinks

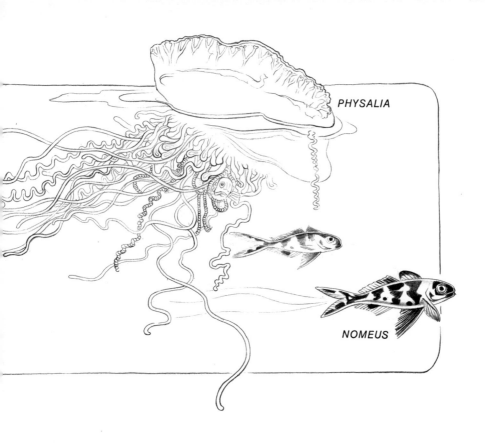

PHYSALIA

NOMEUS

that perhaps the hatchling turtles swim here from the beaches where they are born. Sargassum could conceal these vulnerable infants, which are only the size of a twenty-five-cent coin. The floating weed could also provide these small air-breathing reptiles with creatures tiny enough for them to eat.

The solution to another of nature's mysteries

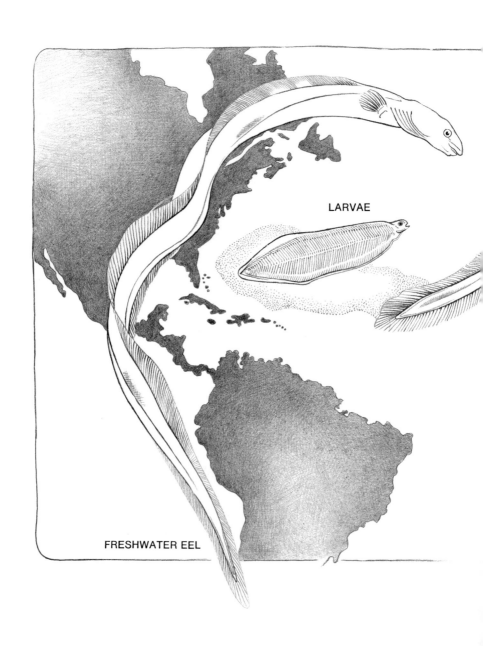

LARVAE

FRESHWATER EEL

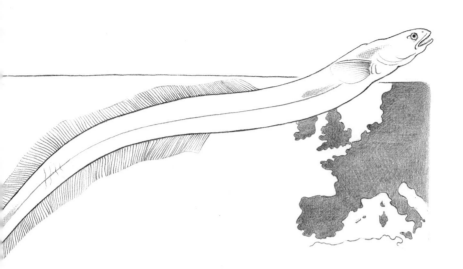

has been solved in the Sargasso Sea. For ages, the spawning ground and migratory patterns of the common North American and European freshwater eels were unknown. The great Danish marine biologist, Johannes Schmidt, took up this investigation. A conscientious and determined scientist, Schmidt spent eighteen years combing the North Atlantic for evidence of the eel's birthplace. Like a patient detective tirelessly gathering clues, he collected the immature young of the eel, transparent larvae resembling willow leaves. Finally

Schmidt found the spawning ground near Bermuda in the Sargasso Sea. It is believed that the eel is born here and migrates to the continents of North America and Europe, where it spends most of its life maturing in freshwater lakes and streams before returning once again to the depths of the Sargasso Sea to spawn. Understanding the life cycle of the eel is important, for the eel is a significant food fish for the peoples of Europe.

VI   A LIVING LABORATORY

In a world faced with increasing shortages of food and other basic resources, marine scientists study the Sargasso Sea. It has become a living laboratory for biologists, ecologists, and oceanographers. Some scientists believe that sargassum, because it is high in protein, could become a valuable food resource. Others think it may have potential use as a fuel. NASA, the

77

National Aeronautics and Space Administration, cooperated with these researchers and attempted to survey the extent of the weed. In 1973, Skylab, the orbiting space station, took photographs of the weed for this purpose during its passes over the Sargasso Sea.

Research vessels gather weed and marine life for other investigations. There is concern about the extent and effect of pollution here. Poisonous pesticides such as DDT have been detected recently in the Sargasso Sea. These chemicals, carried great distances by the winds, settle like dust upon the surface waters and pose a potential threat to marine life. The interaction of pesticides and other pollutants with sargassum and its inhabitants is under study by ecologists. Buttons, jewelry, cigar holders, and other plastic pollutants are now regularly found here also. These items may have traveled vast distances upon the currents. More probably, however, they have been

dumped from passing ships, since the floating debris seems to be concentrated along major shipping lanes.

Ships are also responsible for much of the oil that increasingly pollutes these waters. They rinse their tanks out at sea. The oil is changed in the water into lumps and globs of tar that float. Recently oceanographers, from the Woods Hole Oceanographic Institution searching the Sargasso for weed specimens, had their work interrupted repeatedly because oily tar clogged their nets. Some of these globs were two inches thick. Three times more tar was taken in places than weed. The Skylab photographs may help measure the extent of oil pollution in the Sargasso. It is feared that continued pollution may endanger life in this fragile, unique, and ancient community. Man may in a short time destroy a community that took nature hundreds of thousands of years to develop. Distant as the Sargasso Sea is, it is not remote enough to escape man's reckless pollution.

TAR-CLOGGED NETS HAMPER
THE SEARCH FOR WEED SPECIMENS.

Another area of research concerns weather patterns. The Sargasso Sea may profoundly affect the climate of northwestern Europe. Contrary to the popular and long-held view that the Gulf Stream is primarily responsible for Europe's mild climate, Doctor Henry Stommel at Woods Hole, and other scientists, too, now credit the Sargasso Sea with the major role. They believe that northern Europe is actually warmest when the Gulf Stream and North Atlantic Current are at their weakest. At such times, the Sargasso Sea moves northward into a position where prevailing westerlies cross it.

There winds that blow from west to east, for winds are named by the direction from which they come, absorb the Sargasso's heat and carry it to Europe. It may well be that the Sargasso Sea accounts for the rose gardens of England, the vineyards and wheat fields of France, the ice-free ports of Norway, and the amazing fact that tropical palm trees grow in Ireland, a country so far to the north. That north-western Europe is not a barren wasteland may actually be due to the remarkable Sargasso Sea.

The Sargasso is also the source of weather in the North Atlantic, affecting the ocean and the continents that border it. Winds and currents here create unstable conditions that give birth at times to storms, gales, and destructive hurricanes. For these reasons, the United States Naval Weather Service Command in Washington, D.C., constantly monitors conditions in the Sargasso Sea. Data from orbiting satellites, ships, planes, and weather stations is gathered and analyzed. Patterns and changes are noted. This vital information is made available to meteorologists, weather experts. Weather in the Sargasso Sea ultimately affects all activities in the North Atlantic, influencing air travel, shipping, fishing, and naval operations.

MONITORING AT THE EARTH-ORBITING LABORATORY, SKYLAB

The Navy also studies the relationship between atmospheric conditions and currents in the Sargasso Sea. For scientists now know that undersea storms occur and that they touch off storms in the atmosphere or vice versa.

Scientists from many nations work together in the Sargasso Sea. Project MODE, the Mid-Ocean Dynamics Experiment, is a cooperative effort to study deep ocean currents. The site of this project is the western Sargasso Sea in the area known as the Devil's Triangle. Some people believe that these powerful currents can pull large ships down into the sea. Project-MODE scientists may learn whether this theory is possible. They may also help to solve the mystery of the Devil's Triangle.

VII THE MYSTERY OF THE DEVIL'S TRIANGLE

No ancient legend of the Sargasso Sea is more chilling, more frightening, or more puzzling than the mystery of the Devil's Triangle. This area is where planes, ships, and people have disappeared. They have vanished here without a trace, their fate unknown. The Devil's Triangle, which is also called the Bermuda Triangle, is not shown on any map. But you can

87

BRITISH AIRPLANE (1948)

BERMUDA

S.S. *SANDRA* (1950)

FIVE NAVY TORPEDO BOMBERS (1945)

MIAMI

DEVIL'S, OR BERMUDA,
TRIANGLE

BAHAMAS

CUBA

HAITI

PUERTO RICO

DOMINICAN
REPUBLIC

locate it. Draw a line between Bermuda and
Miami, Florida. From this point, extend a line
southeastward to Puerto Rico, and then con-

nect this second point back to Bermuda, and you will have outlined the area of the Devil's Triangle.

On December 5, 1945, at 2:10 P.M., five Navy torpedo bombers with fourteen airmen left the Naval Air Station at Fort Lauderdale, Florida, for a routine training flight out over the Atlantic. Its course would take them into the area of the Devil's Triangle. Weather conditions were fine for flying. At about 4 P.M. the radio man in the Fort Lauderdale Air Station began to receive alarming messages from the flight patrol leader.

"Everything is wrong . . . strange. We can't be sure of any direction. Even the sea doesn't look as it should."

Only silence followed. At 4:25 P.M., a special rescue plane able to land at sea hurriedly took off in search of the missing patrol. It, too, vanished. A massive air and sea search failed to find any trace of the missing aircraft.

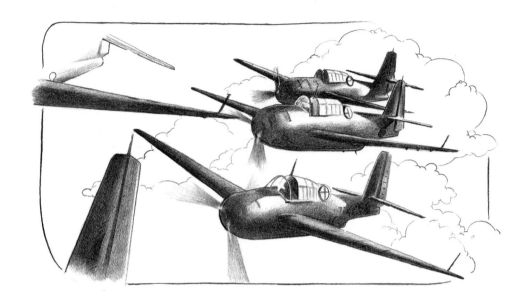

A Naval Board of Inquiry later investigated the disaster and reported: "We are not able to make even a good guess as to what happened."

In January, 1948, a four-engined British airplane approaching an airport in Bermuda on a landing path over the Triangle disappeared with twenty-seven persons aboard. Months later another British aircraft with twenty-four on board left Bermuda en route to Kingston, Jamaica, and was never seen again. No distress calls were heard from either plane.

In June, 1950, the S. S. *Sandra*, a freighter some three hundred and fifty feet long, carrying a cargo of insecticide, left Savannah, Georgia, destined for Puerto Cabello, Venezuela, and vanished somewhere off the coast of Florida.

What happened to these and other ships and airplanes that have disappeared here? Why were they lost? Why were they unable to communicate? Why have no wreckage, debris, or other clues to these disasters been found? Wild and imaginative theories have been offered as strange as the legends of old. Unidentified Flying Objects, UFO's, bearing invaders from outer space have been blamed for the disasters. Others speculate that the vessels entered a crack or hole in space, a so-called space warp.

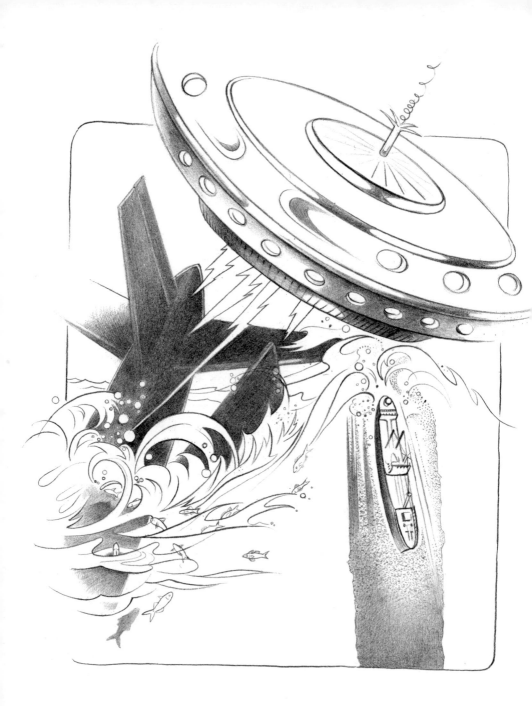

The true cause of these tragedies is still unknown. Certainly natural events of great force do affect a particular area for only a brief time. An earthquake, for example, lasting only seconds, can destroy part of a city leaving other sections undisturbed. Although such an occurrence on land would leave abundant evidence and many witnesses, at sea a great natural event might leave few traces.

In the Triangle, sudden storms, violent air movements, and water spouts are known hazards. Magnetic storms, powerful bursts of energy from the sun, periodically disturb the earth's magnetic field interfering with radio signals and navigational instruments. They also cause unusual atmospheric conditions. No conclusive evidence, however, has proven that any of these natural causes is responsible for these disasters. Scientific investigation thus far has not found anything unusual about the Triangle to account for them.

The riddle of the Devil's Triangle contributes to the legend and mystery that have always surrounded the Sargasso Sea. The efforts of many dedicated men and women of science, however, are revealing the true wonders of this remarkable ocean desert.

INDEX

95